A Mathematical Gallery

A Mathematical Gallery
Lisl Gaal

AMERICAN MATHEMATICAL SOCIETY
Providence, Rhode Island

2010 *Mathematics Subject Classification.* Primary 97B20.

For additional information and updates on this book, visit
www.ams.org/bookpages/mbk-111

Library of Congress Cataloging-in-Publication Data
Names: Gaal, Lisl, 1924- author.
Title: A mathematical gallery / Lisl Gaal.
Description: Providence, Rhode Island : American Mathematical Society, [2017]
Identifiers: LCCN 2017027805 | ISBN 9781470441593 (alk. paper)
Subjects: LCSH: Mathematics – Popular works. | Mathematics – Pictorial works. | AMS: Mathematics education – Educational policy and systems – General education. msc
Classification: LCC QA93 .G225 2017 | DDC 510–dc23 LC record available at https://lccn. loc.gov/2017027805

Copying and reprinting. Individual readers of this publication, and nonprofit libraries acting for them, are permitted to make fair use of the material, such as to copy select pages for use in teaching or research. Permission is granted to quote brief passages from this publication in reviews, provided the customary acknowledgment of the source is given.

Republication, systematic copying, or multiple reproduction of any material in this publication is permitted only under license from the American Mathematical Society. Permissions to reuse portions of AMS publication content are handled by Copyright Clearance Center's RightsLink® service. For more information, please visit: **http://www.ams.org/rightslink**.

Send requests for translation rights and licensed reprints to **reprint-permission@ams.org**.

Excluded from these provisions is material for which the author holds copyright. In such cases, requests for permission to reuse or reprint material should be addressed directly to the author(s). Copyright ownership is indicated on the copyright page, or on the lower right-hand corner of the first page of each article within proceedings volumes.

© 2017 By the author. All rights reserved.
Printed in the United States of America.

∞ The paper used in this book is acid-free and falls within the guidelines
established to ensure permanence and durability.
Visit the AMS home page at **http://www.ams.org/**

10 9 8 7 6 5 4 3 2 1   22 21 20 19 18 17

Contents

Editor's Preface	vii
Author's Preface	ix
1. Counting	1
2. The Pythagorean Theorem	5
3. The Volume of a Pyramid	9
4. The Area of a Circle	13
5. Archimedes's Proof for the Volume of a Sphere	17
6. Pascal's Triangle	21
7. Gaussian Integers	25
8. Permutations	29
9. The Mathematics of Probability and Markov Processes	33
10. Desargue's Theorem	37
11. Seven Circle Theorem	41
12. Calculus	43
13. Multiplying Ordinals	47
A Daughter's Perspective	51
About the Author	53

Editor's Preface

It is said that the best mathematics has within it a simplicity that appeals to the child in us. But while mathematical ideas may have intuitive and direct impact, explaining mathematics fully requires an arsenal of precise terms and rigorous statements that can seem off-putting to the casual observer.

The author uses her own drawings of everyday scenes to gently break through the barriers of language and capture the essence of mathematical ideas from counting to calculus. Originally conceived as a gift from the author to her family, this book is equally accessible to the casually curious and the budding mathematician.

Like a patchwork quilt, the pages of this book do not present a single story, but rather Lisl Gaal's personally chosen sampling, meant for the reader to contemplate and enjoy at leisure.

■ ■ ■

For readers who would like to explore further, comments and suggested projects and exercises have been added to the original text at the end of each section.

The supplemental sections of the text owe a great deal to Marjorie Sayer, for her indispensable insight and advice, and to Marcia Almeida, Barbara Gaal, Meaghan Healy, Doria Hughes, Mark, Ella and Alex Russell and Curtis McMullen for their time, helpful comments and support.

Eriko Hironaka
American Mathematical Society

Author's Preface

Why did I make a mathematical gallery? Lithography combined with watercolor has been my hobby for many years and mathematics my profession, so sometimes, mathematics creeps into my pictures, though it is not always easy to see.

This essay by my granddaughter, Joanna R. Ingebritsen, written when she was 11 years old, expresses more clearly how mathematics and the senses combine.

Math and Me

1. Color: If math were a color, it would be a bright purple. Math is so up and down. One moment you are happy with the way things are going, the next you're frustrated. Purple is composed of two colors that describe this experience. Red describes excitement, blue symbolizes contentment, and purple is made up of both of these.

2. Sound: If math were a sound, it would be a soft, constant hum. Math is everywhere, though it wouldn't be too noticeable for you don't ever notice it unless you think about it.

3. Taste: If math were a taste, it would be the taste of soda. Soda has a taste really different from anything else. The fizziness seems to give it an attitude: explosive and fun at the same time. Sometimes math can be fun; other times it makes you cranky and you feel you're going to explode with anger.

4. Emotion: If math were an emotion, it would be mischievous. You never know what you can do with it or what it might lead to. It can cause lots of trouble, too, because so many people have trouble learning a new concept, or figuring out a problem.

5. Texture: If math had a texture, it would be rough. Here and there are little imperfections called: taking too long and hard to grasp.

All in all, math could be many things: Purple, a constant hum, the taste of soda, mischievous and rough. But, most of all the two words that describe math are: useful and great.

Mathematics can also be very beautiful when a hard problem turns out to have a very elegant solution. I hope that these pictures give you a little insight. If the mathematics at any point looks too hard, just skip it, or leave it for later.

1. Counting

The picture on the left shows that when you count something, you must be very careful to specify exactly WHAT you are counting.

Frequently people see five cats immediately, others require more time. The fifth sits between the two cats on the right.

As the picture of cats shows, a counting problem can only be answered with certainty if there is no ambiguity about which objects you are counting.

There is a whole field of mathematics focusing on problems that involve counting. It is called combinatorics.

■ ■ ■

In counting the cats, do we know if the fifth cat should be counted or not? Do we mean the outline of the cat, or the image that the outline suggests? Could one say that there are no cats in the picture?

Sometimes it is only by finding a careful way to state a problem that one can find an answer. It can also happen that in the process of stating a problem precisely, one finds hidden nuances and complexities.

Here is another example. How many circles are there in the two pictures below?

 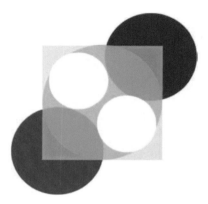

If you look carefully, you will see that the two pictures are exactly the same except for color and layering.

Sometimes with the right formulations, one can find an underlying essence and simplicity.

For example, whether or not you believe there is a fifth cat, you may convince yourself that there are two clusters of cats, one with two cats and one with two or maybe three.

The two figures below are made up entirely of circles. How many are in each? It may help to see how the two pictures are related, and to know that the blue circles come in clusters of seven.

Can you think of a seemingly easy counting problem that is surprisingly hard to state precisely?

Here are just a few random samples:
- How many planets are there?
- Given three points on a plane, how many lines pass through at least two of the points?
- How many different ways are there to arrange five objects in a row?

What are the ambiguities in these questions that can lead to more than one answer? How can you make the questions more precise?

2. The Pythagorean Theorem

The Pythagorean Theorem, named after Pythagoras of Samos (c. 570-495 BC), states that the square of the hypotenuse equals the sum of the squares of the remaining sides of a right triangle.

The picture on the left is a familiar proof of the Pythagorean Theorem using squares with sides of length $a + b$. You can "see" the proof by imagining the triangles moving around and rearranging themselves while keeping their shape, like the wings of butterflies.

The dog on the left of the picture looks quite uncomfortable. By shifting the triangles around, the dog has found a more comfortable position. Compare the sizes of the squares that make up the rest of the picture.

This should lead you to the famous Pythagorean equation
$$a^2 + b^2 = c^2.$$

If you prefer, you can think of the shapes as pieces in a puzzle that can be put together in more than one way, as on the next page. Like the cats and dogs in the picture, you may feel bewildered at first, but the labels should help.

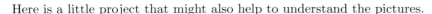

Here is a little project that might also help to understand the pictures.

Say you have a right triangle with side lengths a, b and c.

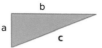

Take two equal squares, and scale your measurements so that the sides of the squares have lengths $a+b$.

See what happens when you cut out the triangles in the two ways shown. For each you should get four copies of the triangle above, and one or two squares left over. Since the triangles are the same size, the areas of the leftover squares must add up accordingly, yielding the equation: $a^2 + b^2 = c^2$.

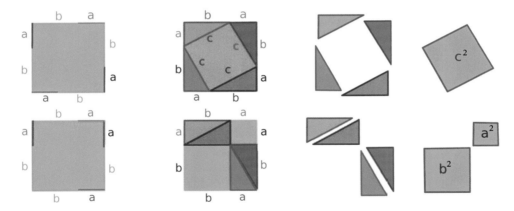

By the way, since we have decomposed the square of sides $a+b$ in this way, we can also derive the area T of the right triangle with sides a, b and c just knowing that the area of a square is the square of its side lengths:

$$a^2 + b^2 + 4T = (a+b)^2 \qquad \text{so} \qquad T = \frac{(a+b)^2 - a^2 - b^2}{4} = \frac{2ab}{4} = \frac{ab}{2}.$$

Two cats and three pyramids make a cube   Lisl Gaal 2009

3. The Volume of a Pyramid

Going further back in time, consider the Great Pyramid of Giza in ancient Egypt (c. 2500 BC). We know it is huge, but what exactly is its volume? Can we find it using outside measurements like sides of the base, and the height?

In the facing picture, cats are playing with smaller pyramids, showing them from different angles.

If the height of the pyramid equals the length of each side of the square base, then we can arrange three such pyramids to form a cube. This means that the volume of the pyramid is one third of the volume of the cube!

You can do this without the help of the cats, but it is not as much fun.

■ ■ ■

This is yet another example of cutting and rearranging to find out about sizes of shapes.

The pyramids we are thinking of have tops that lie directly over one of the corners of the base. It is this shifted version of the pyramid, or *right-angled pyramid*, that fits together with two others to form a cube.
Here is a close-up of how the right-angled pyramids fit together. You can also try making a cube out of jelly and cutting it into three.

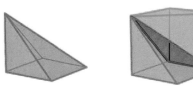 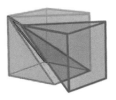

There are two ideas at work here. One is that the volume of a pyramid only depends on the area of the base and the height. The other is that certain pyramids can be fit together to form a cube.

You can see the same ideas in a dimension lower. As you know, the area of a triangle is half of the height times the base, which is the same as half the area of the smallest rectangle that contains it. For the red triangle this is easy to see. What about for the blue ones?

What does this say about the Great Pyramid? Imagine cutting it into four right-angled pyramids as in the pictures below.

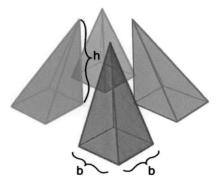 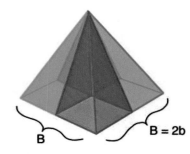

As we see, the Great Pyramid has base equal to four times the area of the four right-angled ones. Each of the four right-angled pyramids has volume equal to one third the base times height. Adding up and using the distributive law we see that the Great Pyramid also has volume equal to one third of *its* base times height.

$$\text{volume of 4 right-angled pyramids} = \frac{b^2h}{3} + \frac{b^2h}{3} + \frac{b^2h}{3} + \frac{b^2h}{3} = \frac{4b^2h}{3} = \frac{B^2h}{3} = \text{volume of Great Pyramid}$$

Visualizing and drawing a right-angled pyramid is a bit tricky. Making a paper model may help.

Paper Model: The faces of the pyramid will have five pieces: the square base, and four triangles. The figure on the left shows how the sides of an Egyptian pyramid might look (scaled down of course!). The one on the right shows the sides of a right-angled pyramid. The bottom is a square, two of the adjacent sides are isosceles right triangles, and the other two are triangles whose sides have ratios $[1 : \sqrt{2} : \sqrt{3}]$. (Some might think these shapes look like flowers. Look carefully at the first picture of this section.)

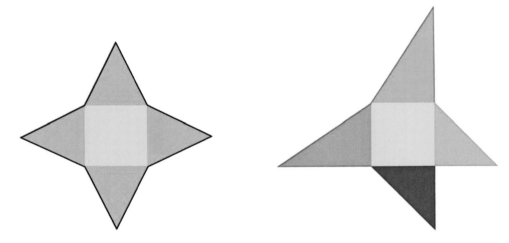

Trace out the shape on the right and make three copies. By cutting, folding and taping you can make three pyramids. Now put the three together to form a cube!

4. The Area of a Circle

The ancient Greeks already knew that there was a special number which always told them the length of the circumference of a circle with radius r whenever this special number was multiplied by $2r$. This special number is called π.

Now what about the area inside the circle? Here's one way the familiar formula for the area inside a circle may have been discovered.

Imagine that the inside of the circle at the top of the plant is covered by green string arranged in concentric circles. The little critters cut and unravel the string and lay them in straight lines. The strings together now form a triangle with base length equal to the circumference of the circle and height r.

Remember that the area of a triangle is always half the length of the base times the height.

In this case, the triangle that we get will have base $2\pi r$, the circumference of the largest circle, and the height is r. It follows that the area A inside the circle is given by:

$$A = \frac{1}{2}(2\pi r)r = \pi r^2.$$

This verifies the well-known formula for the area inside a circle of radius r.

What is going on in this method? How generally does it work?

Let's try some other shapes.

Arrange the string in concentric squares. For a square with sides of length s, the circumference is $4s$ and the shortest distance to the center is $\frac{1}{2}s$. Cut and spread out the strings as before. The triangle you get will have base length equal to $4s$ and height $\frac{1}{2}s$. The area of the middle triangle and the area of the square are equal:

$$\frac{1}{2}(4s)\frac{1}{2}s = s^2.$$

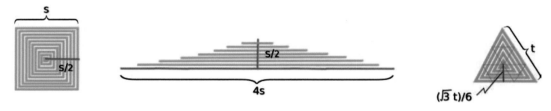

We can try the same method for the equilateral triangle on the right to get the area $\frac{\sqrt{3}t^2}{4}$.

BUT, there is something strange here. Can you spot it?

Finding areas using strings to measure circumferences has its limitations, which you may already be wondering about. How did we choose the distance of the outer string to the center, which became the "height" of the triangle for our calculations? For a circle, there is really only one natural choice, the radius, but for the square or equilateral triangle, there are several.

Consider using the diagonal distance to the outer square instead of the perpendicular one. Then the triangle formula gives the incorrect answer of $\sqrt{2}s^2$ for the area.

What is at work here is not just the length of the string, but also more subtle properties of shape and choices of measurements.

For example, the bowl-like shape below (shown in side view and top view) is also made up of concentric circles, but different techniques (such as calculus) are needed to find its area.

Looking at the top view, we can see that the bowl could also be covered with circular loops of yarn. But how the length of string is changing with respect to distance from the origin is different from when we were studying the inside of a flat circle. If you have ever made a coil bowl out of clay or crocheted a hat, you will have experienced this firsthand. The *curvature* of the surface plays an important role.

The way the size of a region grows relative to its boundary reflects the underlying geometry of the space that the region represents. The relationship between size and boundary is sometimes called the *isoperimetric problem*.

5. Archimedes's Proof for the Volume of a Sphere

The picture on the left shows Archimedes hard at work investigating the relations between shapes. Let's "see" how he finds the volume of a sphere.

In the picture, Archimedes has five cylindrical receptacles of radius r. The left one contains a stack of yellow plates of the same radius and thickness. The furry creature on the left hands each plate to Archimedes who punches out a hole from the middle. The holes get bigger and bigger each time. In fact, the radius of the hole equals the height of the plate from the bottom of the middle three cylinders.

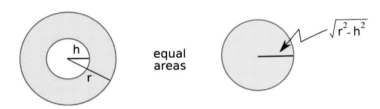

Meanwhile, the fifth, or right-most receptacle contains an upside down hemisphere also sliced horizontally to make disks of larger and larger radius. The furry creature on the right hands these disks to Archimedes, who compares their weights to the weights of the washer shapes, and finds ... they are the same!

Using this discovery and the known volumes of a cylinder and a cone, Archimedes is able to derive the volume of the sphere:

$$\frac{4}{3}\pi r^3.$$

Archimedes's solution to finding the volume of a sphere illustrates the power of using algebraic equations and geometry together.

Let's blow up part of the picture on the previous page to see better what Archimedes is doing. We don't have to weigh the washers and plates to compare their volumes, we can do it mathematically by comparing the areas of the horizontal slices.

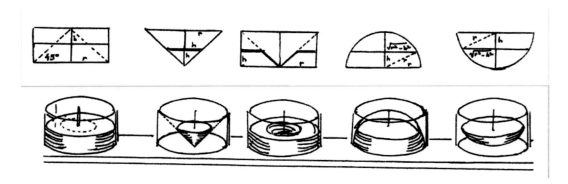

At each height h, the washer has outer radius r and inner radius h (as in the top picture second from left). Using the formula for the area of a circle, it follows that
$$\text{area of washer slice at height } h = \pi r^2 - \pi h^2.$$
As for the plate, the Pythagorean equation tells us that when the height is h, the radius is $\sqrt{r^2 - h^2}$, and
$$\text{area of hemisphere slice at height } h = \pi(\sqrt{r^2 - h^2})^2 = \pi r^2 - \pi h^2.$$
So the area of the washer and the area of the hemisphere slice are equal.

These calculations show that if Archimedes had actually done the experiment shown in the picture, the weights of the washer and plate would be the same as long as the thickness and the height from the bottom are the same.

By verifying that the corresponding washers and plates have the same weight, Archimedes showed the following surprising relation between volumes:

The volume of a cylinder of radius and height equal to r is πr^3, and the volume of a cone with base radius and height both equal to r is $\frac{1}{3}\pi r^3$. Using these formulas, Archimedes could conclude that the *volume of the sphere* is:

$$\frac{4}{3}\pi r^3 = 2(\pi r^3 - \frac{1}{3}\pi r^3).$$

6. Pascal's Triangle

We now turn to a triangle that comes from polynomial equations in a very curious way.

In France, in the year 1635, the 13-year-old Blaise Pascal discovered a triangle of a new sort. In middle-school or high-school algebra you may have learned that

$$(x+1) = x+1$$
$$(x+1)^2 = x^2 + 2x + 1$$
$$(x+1)^3 = x^3 + 3x^2 + 3x + 1$$

and so on...

If we lay out the coefficients of $(x+1)^n$, we get the following diagram, called Pascal's triangle:

$$1 \quad 1$$
$$1 \quad 2 \quad 1$$
$$1 \quad 3 \quad 3 \quad 1$$
$$1 \quad 4 \quad 6 \quad 4 \quad 1$$
$$1 \quad 5 \quad 10 \quad 10 \quad 5 \quad 1$$
$$1 \quad 6 \quad 15 \quad 20 \quad 15 \quad 6 \quad 1$$

and so on...

In this form, we see that in each row after the first, the beginning and end number is 1, and each of the other numbers is the sum of the two numbers directly above it. Try adding up the numbers in each row. Will we always get a power of 2? One way to see why is to plug the value $x = 1$ into the equation $(x+1)^n$.

In the picture, the farmer of Pascal's gardens is still holding one of the pumpkins from the first row. I think he did not make the top shelf quite wide enough.

There are lots of different ways to think about how Pascal's triangle works. Some people may like the algebraic way shown on the previous page. It comes down to seeing the number of ways that a power of x appears in the expansion of $(1+x)^n$, for $n = 1, 2, 3, \ldots$.

Here is another way to look at the numbers in Pascal's triangle.

Imagine Pascal's triangle as paths going down a hill. At each crossing, there are two choices for how to continue downward.

A red and and a blue path are marked on the diagram below, starting at the top of the triangle, marked **start**, and going downward toward the crossing marked **end**. Can you see that there are 8 other paths from the **start** to the **end**? In fact, the numbers in Pascal's triangle tell you exactly how many paths there are from the top of the triangle to the position of the number.

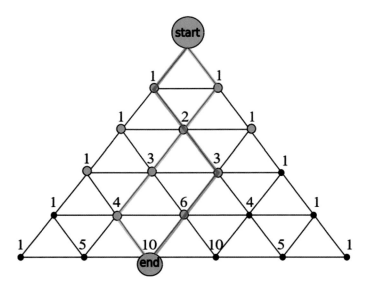

Thinking about path counting may help you to see why each entry of Pascal's triangle is the sum of the entries diagonally above. Why are there '1' s going down both sides of the triangle? Do the rows read the same way backwards as forwards? Why?

Here is a third way that Pascal's triangle is commonly used.

Consider 5 balls numbered $1, 2, 3, 4, 5$.

How many ways are there of choosing 2 of the 5 balls?

You can think of Pascal's triangle as a decision tree. As you go down the paths, you have two choices at each crossing. Think of each descent from one row to another as a choice made. First you choose whether to take ball 1, next whether to take ball 2, and so on. If choosing the right path means "yes" and choosing the left path means "no", then the red path is the choice of balls 2 and 3 and the blue path is the choice of 1 and 5.

It follows that the number of ways to choose 2 out of 5 balls (or *5 choose 2*) is the same as the number of ways to choose paths going downward from the top of Pascal's triangle to the crossing marked **end**. So 5 choose 2 equals 10.

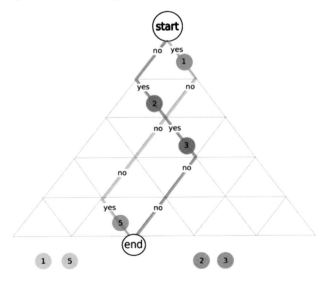

In general to find n choose k, you go to the nth row down from the start, and move k crossings to the right from the leftmost crossing. Can you list the 6 ways of choosing 2 balls out of 4?

the checkered tablecloth   Lisl Gaal ©2002

7. Gaussian Integers

The numbers of the form $a+b\mathbf{i}$, where a and b are positive or negative integers and $\mathbf{i}^2 = -1$, are called Gaussian integers. These play an analogous role in the complex numbers as regular integers do in the real numbers.

The numbers $1, -1, \mathbf{i}$ and $-\mathbf{i}$, are called Gaussian units. A non-unit Gaussian integer which cannot be factored into two other non-unit Gaussian integers is a Gaussian prime. For example, $3, -3, 3\mathbf{i}$ and $-3\mathbf{i}$ are all Gaussian primes, but $2 = (1+\mathbf{i})(1-\mathbf{i})$, so 2 is not a Gaussian prime.

A Gaussian integer which is neither a Gaussian unit nor a Gaussian prime is a composite Gaussian integer.

On the tablecloth in the picnic picture the number $a+b\,\mathbf{i}$ is represented by a square in the (a,b) position. Of course, a and b may be negative. Units are blue, primes are white, composite numbers are red. The little square with 0 in it is just zero.

Tablecloths like this one were on sale in 1954 during the International Mathematical Congresses, a large meeting of mathematicians from around the world that is held every four years.

The food at this picnic is not yet ready, but I think the dog already smells something interesting.

In the realm of *complex numbers*, Gaussian integers play a similar role as regular integers do in the real numbers. For example, we are used to lining up numbers on a number line:
$$\cdots -5, -4, -3, -2, -1, 0, 1, 2, 3, 4, 5 \cdots$$
but why not study a different kind of number system, one that sits on a grid? The Gaussian integers are one such number system.

Addition and multiplication of Gaussian integers are defined using the associative and distributive laws of regular integer arithmetic. For example,
$$(5 + 3\mathbf{i}) + (-2 + \mathbf{i}) = 3 + 4\mathbf{i} \qquad \text{and} \qquad (3 + \mathbf{i})(2 - \mathbf{i}) = 6 + 2\mathbf{i} - 3\mathbf{i} + 1 = 7 - \mathbf{i}.$$

The tablecloth picture depicts the Gaussian integers laid out as squares. Here are the numbers arranged in a grid colored as they are on the tablecloth.

-2+3i	-1+3i	3i	1+3i	2+3i
-2+2i	-1+2i	2i	1+2i	2+2i
-2+i	-1+i	i	1+i	2+i
-2	-1	0	1	2
-2-i	-1-i	-i	1-i	2-i
-2-2i	-1-2i	-2i	1-2i	2-2i

The word unit is used to denote numbers that you can multiply by another to get 1. Since we are thinking of Gaussian integers, the Gaussian units (or simply units) are $1, -1, \mathbf{i}, -\mathbf{i}$, since
$$1 \times 1 = 1, \qquad -1 \times (-1) = 1 \qquad \mathbf{i} \times (-\mathbf{i}) = 1 \qquad (-\mathbf{i}) \times \mathbf{i} = 1.$$

A *composite* Gaussian integer is one that can be written as a product of two numbers neither of which is a Gaussian unit. We saw that 2 is a composite Gaussian integer, since it is a product of $(1+\mathbf{i})$ and $(1-\mathbf{i})$. A number that is not a composite number, and is not a unit, is called a *Gaussian prime*.

On the tablecloth Gaussian units are blue, Gaussian primes are white, composite Gaussian numbers are red. For example, the tablecloth tells us that 5 and $1+3\mathbf{i}$ are composite.

Let's check this:
$$5 = (2+\mathbf{i})(2-\mathbf{i}) \quad \text{and} \quad 1+3\mathbf{i} = (2+\mathbf{i})(1+\mathbf{i}),$$
so both 5 and $1+3\mathbf{i}$ are composite. Can you show that $5+\mathbf{i}$ is composite? How about 13?

Why are the colors symmetric around the "0"?

See what happens when you multiply Gaussian integers by the units. Of course, multiplication by 1 makes everything stay the same. Multiplication by -1 flips everything across the vertical axis (also known as the i-axis). Multiplication by $\mathbf{i}$ rotates the Gaussisn integers counter-clockwise by $90°$ around 0, and multiplication by $-\mathbf{i}$ rotates clockwise by $90°$. Multiplication by units preserves the property of being a Gaussian unit or a Gaussian prime. All these put together show that rotating the tablecloth by $90°$ will not change the pattern.

How are the white and red checks distributed as we go further and further out in the grid?

As in the study of regular integer primes, how Gaussian primes are distributed in the Gaussian integers is a deep and interesting question that has fascinated mathematicians for centuries.

8. Permutations

A permutation of a finite set of objects is a rearrangement of the objects.

The different permutations of the three colors red, green and blue are illustrated in the facing picture. See if you can find all six. It may be easiest to see this by looking at the bottom rungs.

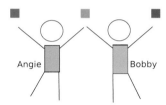

Permutations can be thought of as something you *do* to an arrangement. Like turning the faces of a Rubik's cube to rearrange the colors.

Say that Angie and Bobby have three colored objects (squares) between them. What if whenever you point to Angie she switches the left two objects, and when you point to Bobby he switches the right two. The instructions can be encoded by a list consisting of A's and B's. For example AB is the permutation giving the sequence green, blue, red, while ABA is the permutation giving the sequence blue, green, red.

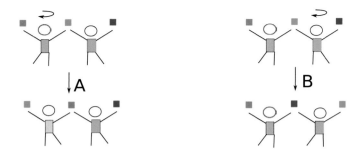

Every permutation of the three squares can be made by a string of As and Bs. There are six distinct ones given by $A, AB, ABA, ABAB, ABABA$ and $ABABAB$. The last one $ABABAB$ is the same as not changing anything.

See if you can verify the following equations and inequality by checking what happens to the sequence of colors.
$$B = ABABA, \qquad ABA = BAB, \qquad ABAB = BA \quad \text{and} \quad AB \neq BA$$
Sometimes drawing a picture of a mathematical idea can suggest a deeper layer behind it. In the pictures you see a progression of permutations, so that at each stage, a pair of colors switches. Such a permutation is called a *transposition*. For example, the actions of Angie (A) and Bobby (B) are transpositions. The pictures show that you can get from any one permutation of three colors to any other by a sequence of transpositions.

The pictures also suggest a relation between permutations and braids. For example, in the second picture you see the three colors as roads that pass over and under each other much like a braid.

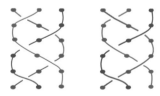

Both of the braids above list our permutations $A, AB, ABA, ABAB, ABABA$ and $ABABAB$, but they are not equal (as anyone who has braided hair knows very well). Braids give an added dimension (technically and figuratively) to the study of permutations.

This picture illustrates the same idea as the first. There are three east-west roads, at various relative distances from the viewer. There are 6 scenes in the picture on this page: New York, New England, Minnesota, the plains, the mountains, and the San Francisco Bay Area. When you look, remember that Minnesota is North of both New York and San Francisco, so you are looking South.

A Minnesotan's View of the USA

The illustration was inspired by Saul Steinberg's March 29, 1976, "View of the World from Ninth Avenue" cover of The New Yorker *Magazine.*

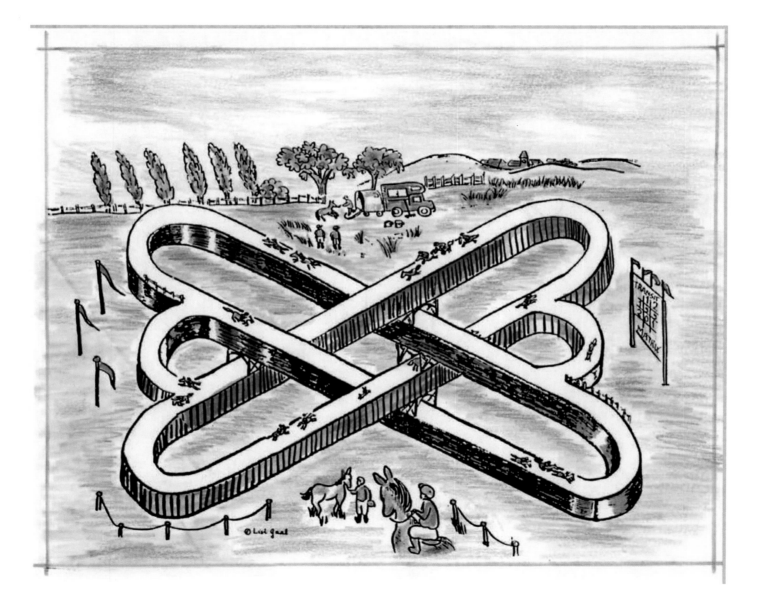

9. The Mathematics of Probability and Markov Processes

The idea of mathematical probability may lead you to think of gambling, and then perhaps to paintings by famous Flemish and Dutch masters of gamblers playing with dice or cards.

But I thought I'd try horses running on a track, and make a picture of a Markov process with two states, each represented by an oval.

What is a Markov process? If you flip a coin over and over, the chance of coming up heads is 50% each time. But the chance of a rainy day occurring right after a sunny one is slightly less likely than a sunny day following another one. On the other hand, perhaps there is a higher chance of rain after a week of sunshine. A Markov process is a sequence of events, called states (like heads or tails, rainy day or sunny day, or a horse running on red oval or blue oval) where each state has a certain probability of turning into another state.

In the picture, all horses start on the blue oval, and each horse has an even chance of staying there or turning to the red oval at the next ramp. Once they change to the red oval, they cannot change back, because you cannot make a galloping horse make a sudden U-turn. Here, the red oval is what is called an absorbing state.

If 20 horses start on the blue oval, what is the probability that all horses are on the red track after three laps?

To solve the problem of what happens at the end of three laps, it helps to begin by finding out the chance of one horse reaching the red track by the third lap. This is because we can assume that whether any particular horse reaches the red track does not depend on what happens to the others. All the horses are governed by the probabilistic model.

If a horse is on the blue oval, then in one lap it will have two chances to switch to the red oval. After half a lap there is an equal chance for the horse in the blue state to either stay in the blue state or change to the red state.

Let's think of the Markov process as a sequence of *stages*. At the first stage (half a lap), a horse has a $0.t$ probability (or 50% chance) of moving to red.

What are the chances after one full lap? These are the possibilities:

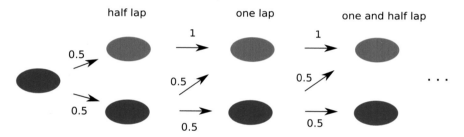

At the second stage (one lap), the probability that the horse is still on the blue oval is a half of a half, or $0.25 = 0.5 \times 0.5$. The probability that the horse is on the red oval is a half plus a half of a half or $0.75 = 0.5 + 0.5 \times 0.5$.

Notice that $0.75 + 0.25 = 1$. This corresponds to the fact that after one lap there are only two possible states for the horse: blue or red. Since there are two possible states at each stage, the probabilities of each state occurring at a given stage always add up to 1.

For each horse, what is the probability that the horse is on the red oval after three laps (that is, six stages)?

At each stage, we multiply the chance of being in the blue oval by a half, and add it to the chance of being in the red oval. This means that at stage six the probability is:
$$0.5 + \left[0.5 + \left[0.5 + \left[0.5 + \left[0.5 + (0.5)^2\right]0.5\right]0.5\right]0.5\right]0.5 = 0.5 + (0.5)^2 + .(0.5)^3 + (0.5)^4 + (0.5)^5 + (0.5)^6.$$

As we saw earlier, we can also solve this problem by finding the probability that the horse stays on the blue oval after six stages. This would mean that at each state, the horse made the choice of keeping to the blue. That is half of a half of a... six times, or $(0.5)^6$ or $\frac{1}{64}$. This means that the probability of the horse to be on the red oval after six stages is $\frac{63}{64} = 0.984375$.

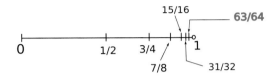

The probability of each horse to be on the red oval after k laps is given by the finite geometric series:
$$0.5 + (0.5)^2 + (0.5)^3 + (0.5)^4 + (0.5)^5 + (0.5)^6 + \cdots + (0.5)^k = 1 - (0.5)^k.$$

Since $(0.5)^k$ gets arbitrarily small as k grows large, we see that the more laps a horse runs, the more likely that horse will end up on the red oval. On the other hand, there is no finite stage where the horse is certain to be on the red oval.

Now, what about the probability of all twenty horses to be on the red oval after three laps? Since the chances of each horse being on the red oval are independent, the answer is approximately $(0.984375)^{20} \approx 0.72981$.

Desargues's Boat

10. Desargue's Theorem

Since we are now on sports, let's consider sailing. This was one of the first math pictures I ever made.

Desargue's Theorem is basic to plane projective geometry, which dates from the seventeenth century when mathematicians started to examine and question Euclid's axioms. In fact, it is one of the axioms of 2-dimensional projective geometry, but oddly enough, it can be proved quite easily in 3-dimensional projective geometry.

Desargue's Theorem says: If two triangles $A_1B_1C_1$ and $A_2B_2C_2$ are such that the lines A_1A_2, B_1B_2 and C_1C_2 all meet in one point, then the points of intersection of their corresponding sides (i.e., the three intersections of A_1B_1 and A_2B_2, of B_1C_1 and B_2C_2 and of A_1C_1 and A_2C_2) all lie on the same straight line.

Let's look closer and extract the lines and triangles making up the sail. In the figure below, the picture on the left is a copy of the one in the picture turned 90° clockwise. The picture on the right shows how the diagram might look if you raised the points B_1 and B_2 up above the plane containing A_1, A_2, C_1, C_2 in 3-dimensional space, while making sure that lines making up the triangles $A_1 B_1 C_1$ and $A_2 B_2 C_2$ remain lines, and the lines through the A's, B's and C's still meet at a point O.

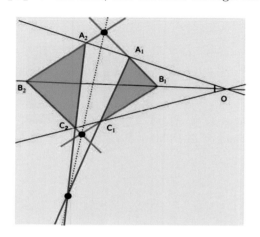 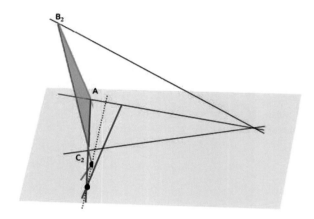

How do we then know that the black dots (denoting the intersections of the pairs of lines AB and BC and AC) will all lie on the same line?

Suppose we only know that the A, B and C lines intersect in a single point O. Consider the two planes P_1 and P_2 containing the original triangles. As long as P_1 and P_2 are not parallel, they must intersect in a line. Let's call that line L. It is the dotted line in the right picture.

Since line containing the A points and the line containing the B points pass through O, the two lines lie on a plane P_{AB}. Similarly there is a plane P_{BC} containing the B and C points, and P_{AC} containing the A and C points. Notice that the two triangles defined by $A_1 B_1 O$ and $A_2 B_2 O$ must both lie on P_{AB}. This means that the extensions of the sides $A_1 B_1$ and $A_2 B_2$ must intersect at a point on P_{AB}. Let's say the intersection occurs at p_{AB}. Similarly, the BC lines and AC lines must intersect at points p_{BC} and p_{AC}. The three points p_{AB}, p_{BC} and p_{CB} have to lie on both P_1 and P_2, so they must lie on their intersection, which is L. All this works no matter how slightly you tip the triangles. This completes the proof of Desargue's Theorem.

The diagram below shows the three dimensional arrangements of triangles and their shadows.

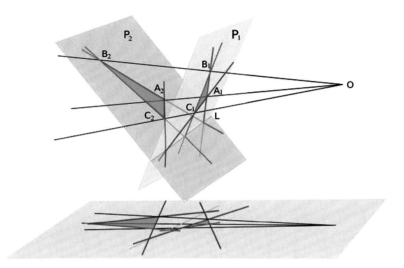

Now that we have proved Desargue's Theorem, what about the converse?

Suppose you have two triangles $A_1B_1C_1$ and $A_2B_2C_2$ on skewed planes P_1 and P_2 in space, and suppose that the pairs of lines AB, BC and AC intersect (and hence lie on the line where the planes P_1 and P_2 meet). Does it follow that there is a point O in space so that the lines through the A, B and C points all pass through O?

The answer again is yes (as long as the triangles are sufficiently skewed). You simply consider the planes defined by AB and BC and AC in 3-dimensional space. Do you see them? Three sufficiently skewed planes must intersect in a point, and that point has to be the desired O.

Just like Desargue's Theorem, the 3-dimensional version of the converse statement also gives a proof for the 2-dimensional version.

11. Seven Circle Theorem

Here is another theorem in plane Euclidean geometry. It was discovered and proved recently by Evelyn, Money-Coutts, and Tyrrell[1]. If a chain of six circles is tangent to a seventh circle, then the three lines joining opposite points of tangency are concurrent.

■ ■ ■

The claim is easy to see if the circles are all the same size. The three lines will then intersect at the center of the seventh circle. But you can also have fun with lots of other arrangements. In the pictures below, the seventh circle is golden, and the other six are in pairs of green, red, and blue.

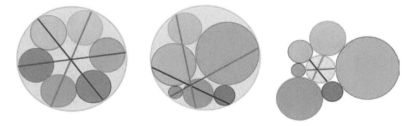

Here you see the circles and circles going round and round...like the merry-go-round of the picture.

---

[1] C. J. A Evelyn, G. B. Money-Coutts, and J. A. Tyrrell, *Seven Circles Theorem: And Other New Theorems*. (London: Stacey International Publishers, 1974)

12. Calculus

Calculus lets you find quantities which are related, i.e., they depend on each other. When driving a car, the speed, acceleration and distance traveled are all related. The speedometer indicates the speed at all times. If you make a graph with time on the x-axis and speed on the y-axis, the slope of the tangent at any point on the graph equals acceleration at that time. The area under the graph over an interval of time will measure the distance traveled during that time. The slope is called the derivative, and the area is called the integral.

When you take a class in calculus, you learn how to calculate these slopes and areas. If you have carefully graphed your function, you can approximate the tangent slopes with an arrow, or approximate the area by filling it with narrow rectangles, but this is a lot of work. When you have a nice formula to describe this curve, the tools of calculus can give you the answer in a jiffy. This is one reason why it is often useful to model complicated information using formulas.

In the picture, the ruler represents the x-axis, the thermometer the y-axis, the edge of the clouds is the function $f(x)$, and the contrail of the airplane is the derivative at the point where it touches the clouds. The outline of the sky-scrapers is an approximation to the area under the curve, i.e., to the value of the integral.

If you have been studying graphs of functions and thinking about derivatives and integrals, you might start seeing city skylines as a step in approximating the integral of the curve defined by some clouds, and you might see a jet stream as part of a tangent line.

Calculus involves geometric understanding and the ability to use and manipulate formulas to make computations. One can get far just knowing computational manipulations, but it always helps to have a *picture*.

For example, tangent lines give an instantaneous picture of how the value of a function is changing at a given moment. If the slope is positive, it means the function is increasing, and the larger the slope, the steeper the rise in value.

The *fundamental theorem of calculus* tells us that if $f(t)$ is the speed of travel, then the area underneath the graph $y = f(t)$ over the interval $[t_0, t_1]$ is the distance traveled between time t_0 and time t_1 – yet another of the surprising relations one finds in mathematics.

Here is a simplified picture of the buildings and clouds, or rather a graph of a positive function over a bounded interval. The integral is the area under the graph and above the interval. Though difficult to do in practice, there is a nice intuitive way to find the area using *Riemann sums*, that is by approximating the area using rectangles.

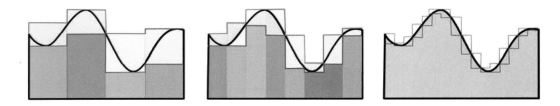

The total area of the darker rectangles and the total area of the light blue rectangles are lower and upper bounds for the area under the curve. Making the widths of the rectangles narrower makes the approximation of the area more accurate.

If the speed of a car is a constant r on the time interval $[t_0, t_1]$, then we know that the distance traveled is $r(t_1 - t_0)$ (speed $\times$ time), or exactly the area under the curve of the graph of $y = r$ over $[t_0, t_1]$. The rectangles in the picture can be thought of as the distance the car would have traveled if its speed were constant over subintervals of a given width (though this is not completely realistic since the change in speed would have to be instantaneous). The distance traveled and the area under the curve are being simultaneously approximated by the same quantities so they must be equal.

13. Multiplying Ordinals

The last picture may remind you of Yosemite or some other beautiful place on earth. It illustrates the question: Is a times b always the same as b times a?

Compare the trees in the foreground of the picture to the waterfalls coming down from the cliffs.

The commutative property holds for finite numbers: the branches of the trees in the foreground show the equality: $2 \times 3 = 3 \times 2$. But it does not hold for infinite ordinal numbers.

What is an ordinal number?

You may say that Maggie has five cats, or you may have listed the cats in order of arrival: first, second, third, fourth, fifth. In the first case, you are dealing with an unordered collection; in the second, the cats are now ordered. Referring to the fifth cat uses the number five in the sense of an ordinal number. Its use in the first case is as a cardinal number.

The smallest infinite ordinal number is denoted by ω (pronounced omega). Thought of as a set ω can be identified with the set of the (ordered) natural numbers $1, 2, 3, \ldots$. You can also visualize ω as a series of telephone poles vanishing into the distance or, as in the picture, as a waterfall.

If a tree first produces two branches and each of these divides into three branches, you get six branches. If it first produces three branches and they each divide into two, you also get six, so $2 \times 3 = 6$ and also $3 \times 2 = 6$. It doesn't matter in what order you count them. This is called the commutative property of multiplication.

Now think of a brook as a single stream of water, and a waterfall as the result of breaking up that stream into an infinite number of rivulets. Now divide each rivulet into two, that is making each rivulet into two rivulets. The waterfall is still a single waterfall. If we do the two steps in the opposite order, we divide the brook into two brooks, and then make each of them into waterfalls. You get two waterfalls rather than one.

This illustrates that $\omega \times 2$ is not the same as $2 \times \omega$!

■ ■ ■

Here is another way to explain the difference between $\omega \times 2$ and $2 \times \omega$ in terms of ordinals.

Let's go back to the finite case. If you see a bunch of people, you can find out how many there are (the cardinality) by counting them starting with 1. The process of counting automatically orders the people, and is useful for determining the cardinality.

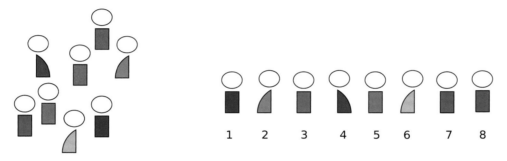

How does multiplication work in the context of cardinals versus ordinals in the finite case? Is there a difference between lining up four pairs of people in a row, and lining up four people and then four more people after them? In both cases you will see a line of eight people.

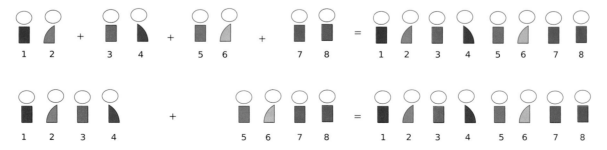

Now picture the ordinal ω as an ordered set of people. We can't see all the people because they are getting further and further away.

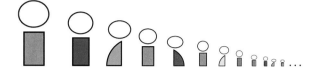

What is $2 \times \omega$? It is two copies of ω one after the other as shown in the figure below.

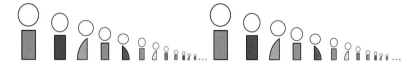

You can see that every person in the second batch has an infinite number of people coming before them. To illustrate, no matter how long they wait, a person handing out water one by one from the beginning would never get to those people in the second batch. Now what about $\omega \times 2$? We can think of this as an infinite number of pairs of people, or perhaps people and their shadows.

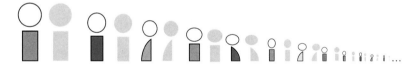

Though it will take longer, the further out you are, a person handing out water will eventually get to each one of the people (and their shadows).

We've seen that $2 \times \omega$ and $\omega \times 2$ are qualitatively quite different from each other. What would $3 \times \omega$ look like? You can draw it with cats, or anything you like. What about $\omega \times \omega$?

Mathematics is all around us, in flowers, ferns, sea-shells, mountains, cities. You just have to look!

I hope you enjoyed my gallery. If there were things that you did not understand fully, set the book down for a while, and come back again later. You may see more than you did the time before.

I want to give Many Thanks to all who encouraged and helped me to write this book and correct my errors: my family, Joanna Ingebritsen, Ellen Ingebritsen, Laura Gould, Helen Lewy, Constance Reid and many others!

A Daughter's Perspective

As a youngster, math was easy for me. My mother Lisl (the author and illustrator of this book) always had time to provide a clear, five-minute explanation and get me back on track. I took this for granted and only realized much later what a gifted teacher she was. Similarly, I did not then appreciate what a brave and talented pioneer she was, breaking barriers as one of the first women on the faculty of the Mathematics Department at the University of Minnesota. In 1952, before my sister and I were born, she had been given one of the 10 "Mlle Merit Awards" by Mademoiselle magazine, along with more famous women like the ballerina Maria Tallchief, actresses Maureen Stapleton and Shelly Winters, and athlete Maureen Connolly. She continually juggled her work responsibilities with the demands of raising my sister and me, at a time when existing systems presented many obstacles, like school lunch-breaks when children were sent home to their presumably stay-at-home moms. Having grown up in Austria, my mother was unfamiliar with American customs, making her multiple roles even more challenging. Throughout it all, we always felt well-loved, and never short-changed.

Mom also maintained an active interest in art, participating in the Art Section of the Faculty Women's Club and doing various creative projects, like helping to design several banners for our local church. After my sister and I grew up, she had time to learn lithography, which became her specialty. Her lively, colorful prints won several prizes at the Minnesota State Fair and were exhibited in many venues.

I always enjoyed her other pictures, but I did not readily relate to her mathematical pictures at first. After she compiled them into a book and added some short and engaging explanatory text, I began to understand them better and appreciate how clever and special they are. It is exciting that her book will now be published and available to a wide audience, thanks to AMS and editors Eriko Hironaka and Marjorie Sayer.

Barbara Gaal